P9-DCD-662

THE SPACE SHUTTLE

by Jeffrey Zuehlke

Lerner Publications Company • Minneapolis

For Wiley, the orbiter

Lerner Publications Company
A division of Lerner Publishing Group
241 First Avenue North
Minneapolis, MN 55401 U.S.A.

Website address: www.lernerbooks.com

Words in **bold** are explained in a glossary on page 30.

Library of Congress Cataloging-in-Publication Data

Zuehlke, Jeffrey, 1968–
The space shuttle / by Jeffrey Zuehlke.
p. cm. – (Pull ahead books)
Includes index.
ISBN-13: 978–0–8225–6420–1 (lib. bdg. : alk. paper)
ISBN-10: 0–8225–6420–3 (lib. bdg. : alk. paper)
1. Space shuttles–Juvenile literature. I. Title.
TL795.515.Z84 2007
629.44'1–dc22 2006018966

Manufactured in the United States of America
1 2 3 4 5 6 – JR – 12 11 10 09 08 07

The countdown is almost finished!
The space shuttle is ready to take off!
Five, four, three, two, one!

Liftoff! The space shuttle is blasting off! It is going on a new trip. A space shuttle trip is called a mission.

Blasting off is the first part of the mission. It is called the **launch.** Up, up, the space shuttle goes! The space shuttle is lifting into space.

The main part of the shuttle is the **orbiter.** It has wings like an airplane. Sometimes people call the orbiter the space shuttle.

The space shuttle has three very powerful rocket engines. They are the main engines.

Two other engines help with the launch. The big, white **solid rocket boosters** push the shuttle into the sky.

See the big orange tank? That is the **external tank.** It is filled with fuel. The engines use the fuel to make power.

Look! The shuttle has dropped the solid rocket boosters!

The boosters have finished their job. They fall into the ocean. A ship comes to pick them up. They will be used for another launch.

The orbiter has used all of the fuel in the external tank. The shuttle drops the tank. It will break apart as it falls back to Earth.

The booster engines have stopped. The shuttle is **orbiting.** It is spinning around Earth.

People who fly the space shuttle are called astronauts. The astronauts fly the shuttle from the flight deck.

Astronauts also work in the shuttle's middeck. People and things float in space. They are weightless.

The astronauts float around the shuttle.
Being in space is sort of like flying!

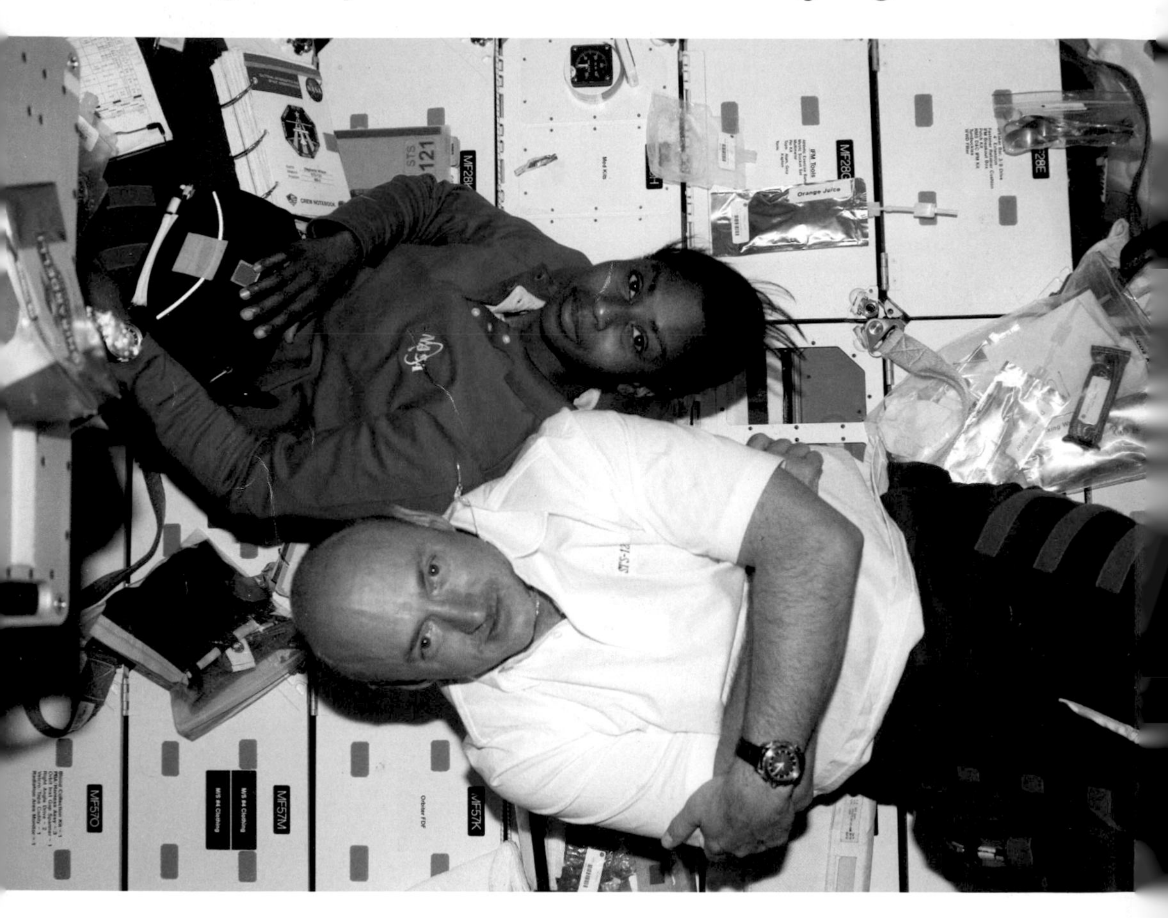

This astronaut is getting dressed up in a space suit. Where is he going?

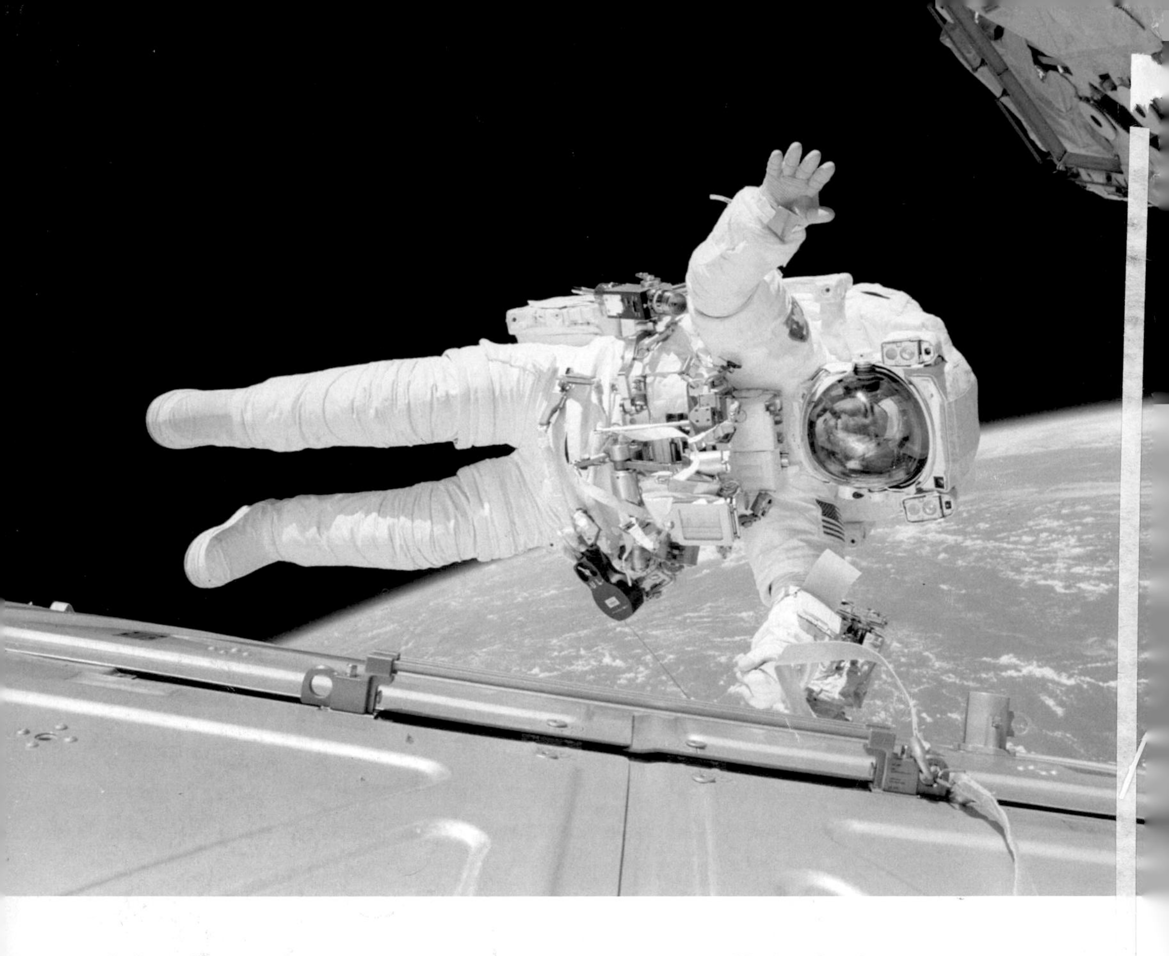

He is going on a space walk! It is very cold in space. But the space suit protects the astronaut.

These astronauts are in the **payload bay.** The shuttle carries supplies in the payload bay. The bay has big doors that open and close.

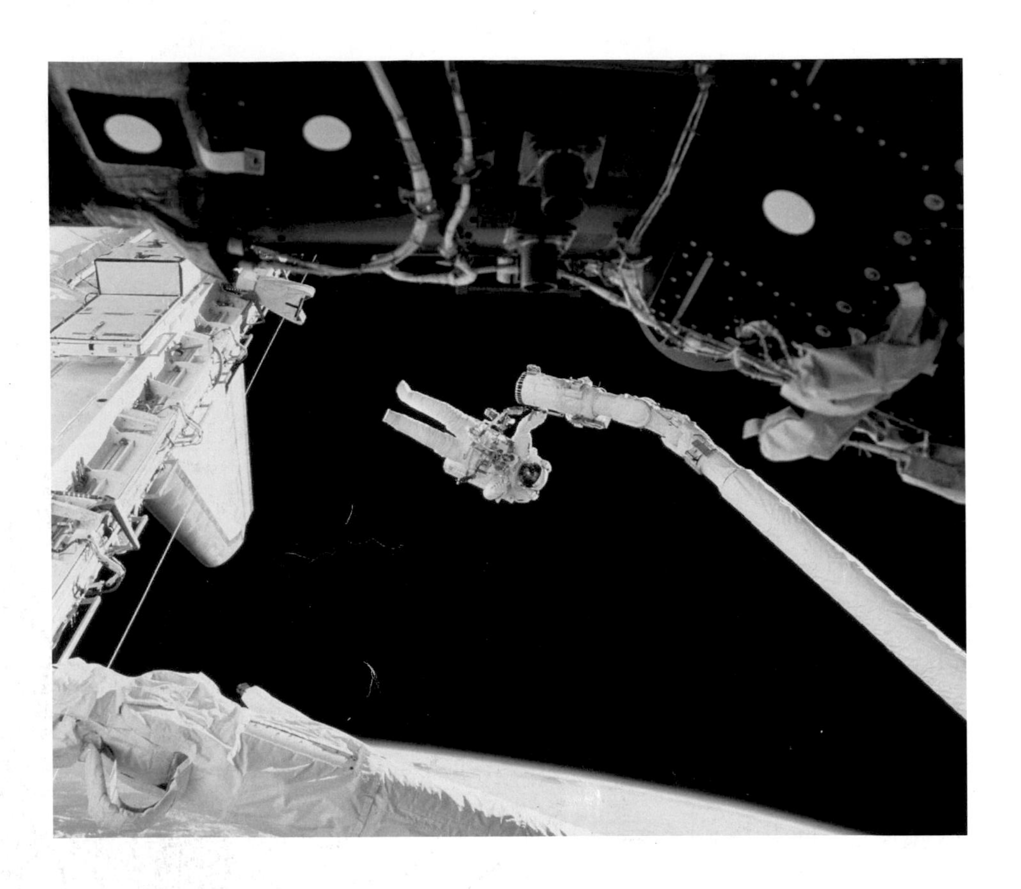

A robot arm is inside the payload bay.
The arm helps the astronauts to lift and
move things.

The arm is lifting a big **satellite.** A satellite is a machine that orbits Earth. The arm helps to put the satellite into space.

Hey! What is that big thing floating in space? It is the International Space Station.

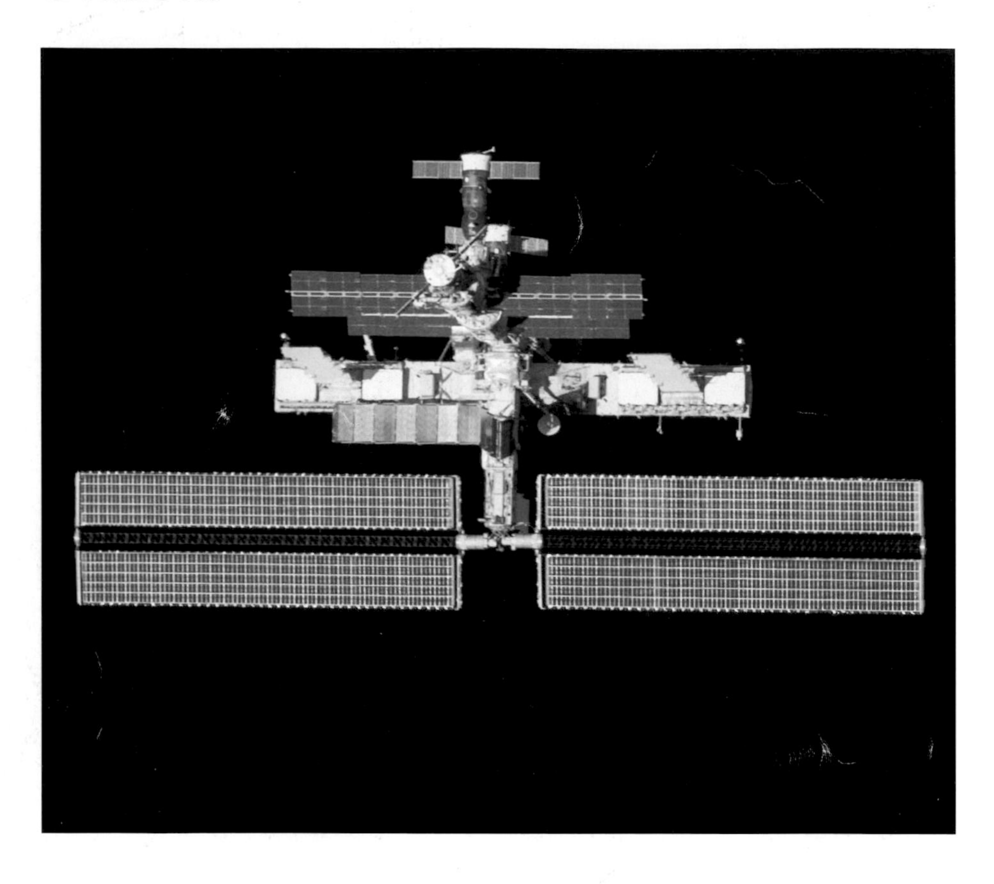

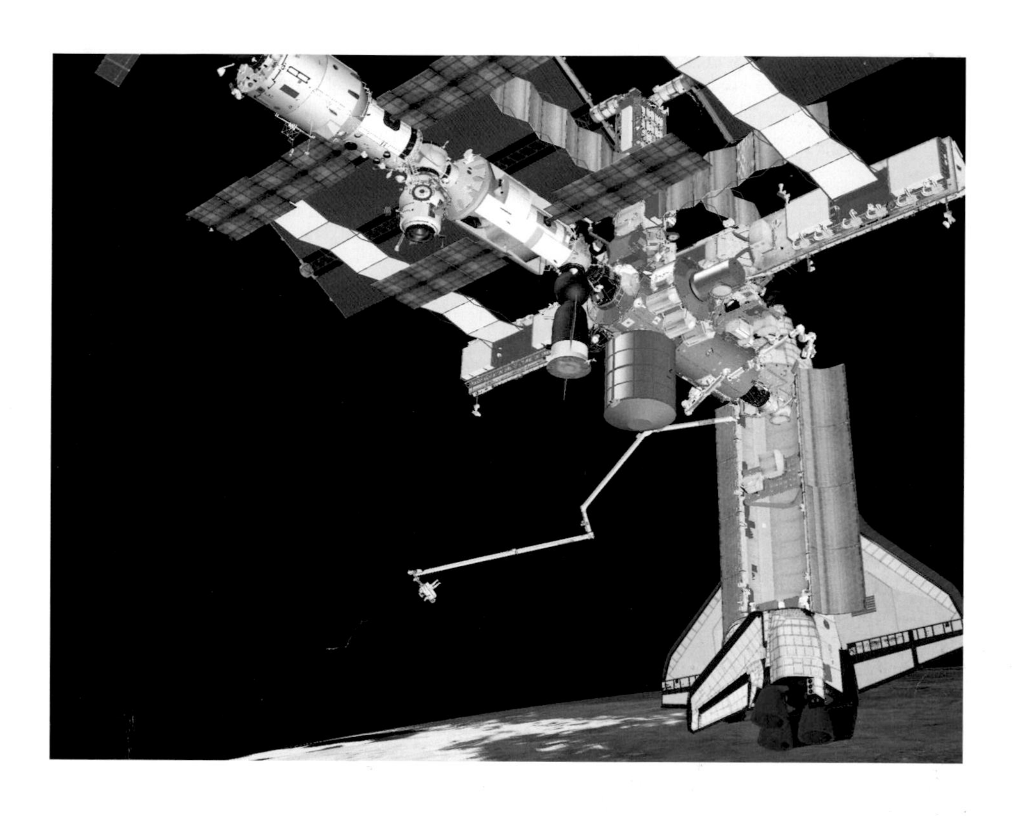

Astronauts live on the International Space Station. The shuttle visits the station every few months. It carries people and supplies to the station.

The shuttle's mission is almost over. It is time to say good-bye and head back to Earth. The return to Earth is called **reentry.**

The shuttle uses its wings to glide to the ground.

The shuttle lands on a runway. A parachute pops out the back to slow the shuttle down.

This airplane is giving the shuttle a piggyback ride. The shuttle is getting a flight back home. It will be ready for the next launch.

Facts about Space Shuttles

- Space shuttle missions are run by the National Aeronautics and Space Administration, or NASA. It is part of the U.S. government.

- NASA has three space shuttle orbiters. Their names are *Atlantis*, *Discovery*, and *Endeavour*.

- The first space shuttle launch took place on April 12, 1981. The name of the first shuttle was *Columbia*.

- Two astronauts flew on the first space shuttle mission. They were John Young and Robert Crippen.

- The space shuttle flies at more than 17,000 miles per hour during launch. It takes the shuttle just eight minutes to reach space.

- Space shuttles have flown more than 120 missions.

Parts of a Space Shuttle

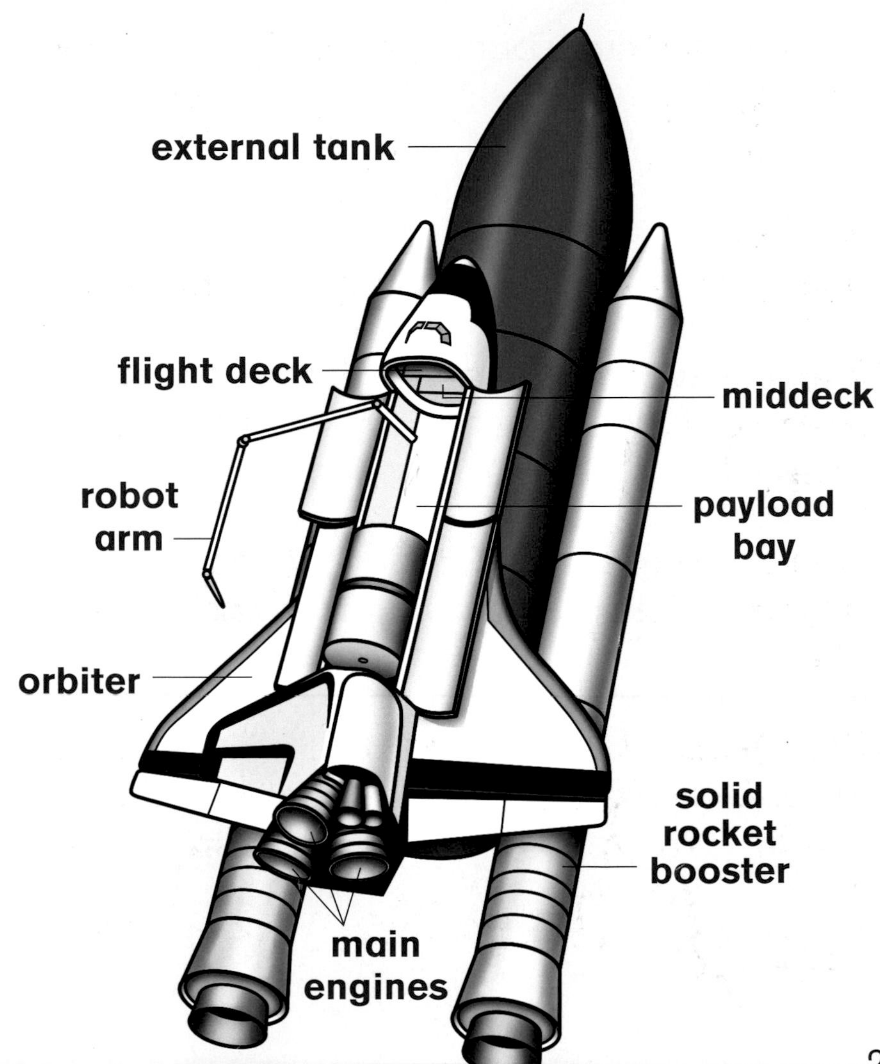

Glossary

external tank: the large tank that holds fuel for the shuttle's launch

launch: blasting off into space

orbiter: the main part of the space shuttle

orbiting: circling around Earth

payload bay: the large area that holds equipment on the shuttle

reentry: returning to Earth from space

satellite: a machine built to orbit Earth

solid rocket boosters: two large rocket engines that help push the orbiter into the sky

More about Space Shuttles

Bredeson, Carmen. *Getting Ready for Space*. New York: Children's Press, 2003.

Bredeson, Carmen. *Liftoff!* New York: Children's Press, 2003.

Feldman, Heather. *Columbia: The First Space Shuttle*. New York: PowerKids Press, 2003.

NASA: Return to Flight Kids' Page
http://www.nasa.gov/audience/forkids/home/returntoflight.html
Visit NASA's website to find activities, photos, and lots of fun facts about the space shuttle.

NASA: Space Place
http://spaceplace.jpl.nasa.gov/en/kids
This NASA website is made just for kids. Check it out to find games, fun facts, projects, activities, and more.

Index

astronauts, 14–15, 16, 17, 18, 19, 20, 23

blasting off, 4, 5

engines, 7, 8, 13

fuel, 9, 12

space station, 22–23
space suit, 17, 18
space walk, 18
supplies, 19, 23

tank, 9, 12

wings, 6, 25

About the Author

Jeffrey Zuehlke remembers watching the very first space shuttle launch in April 1981. He was a school kid living in Saint Paul, Minnesota, at the time. He has watched many launches since then. He hopes to be the first children's book writer to fly on the space shuttle. He hopes NASA will let him bring his son Graham and his dog Wiley with him too.

Photo Acknowledgments

The photographs in this book appear courtesy of: National Aeronautics and Space Administration (NASA), front cover, pp. 3, 4, 5, 6, 7, 8, 9, 11, 12, 13, 14, 15, 16, 17, 18, 19, 20, 21, 22, 23, 24, 25, 26, 27; © Tom Pantages, p. 10. Illustration on p. 29 by Laura Westlund, © Independent Picture Service.